Cambridge Elements ⁼

Elements of Paleontology

FLIPPING THE PALEONTOLOGY CLASSROOM

Benefits, Challenges, and Strategies

Matthew E. Clapham
University of California Santa Cruz

Paleontological
SOCIETY

CAMBRIDGE
UNIVERSITY PRESS

CAMBRIDGE
UNIVERSITY PRESS

University Printing House, Cambridge CB2 8BS, United Kingdom

One Liberty Plaza, 20th Floor, New York, NY 10006, USA

477 Williamstown Road, Port Melbourne, VIC 3207, Australia

314–321, 3rd Floor, Plot 3, Splendor Forum, Jasola District Centre,
New Delhi – 110025, India

79 Anson Road, #06–04/06, Singapore 079906

Cambridge University Press is part of the University of Cambridge.

It furthers the University's mission by disseminating knowledge in the pursuit of
education, learning, and research at the highest international levels of excellence.

www.cambridge.org
Information on this title: www.cambridge.org/9781108717847
DOI: 10.1017/9781108681421

© The Paleontological Society 2018

First published 2018

A catalogue record for this publication is available from the British Library.

ISBN 978-1-108-71784-7 Paperback
ISSN 2517-780X (online)
ISSN 2517-7796 (print)

Flipping the Paleontology Classroom

Benefits, Challenges, and Strategies

Elements of Paleontology

DOI: 10.1017/9781108681421
First published online: October 2018

Matthew E. Clapham
University of California Santa Cruz

Abstract: Lecturing has been a staple of university pedagogy, but a shift is ongoing because of evidence that active engagement with content helps strengthen learning and build more advanced skills. The flipped classroom, which delivers content to students outside of the class meeting, is one approach to maximize time for active learning. The fundamental benefit of a flipped class is that students learn more, but ensuring student preparation and engagement can be challenging. Evaluation policies can provide incentives to guide student effort. Flipping a class requires an initial time commitment, but the workload associated with evaluating student work during the course can be mitigated. The personal interactions from active learning are extremely rewarding for students and instructors, especially when class sizes are small and suitable room layouts are available. Overall, flipping a course doesn't require special training, just a willingness to experiment, reflect, and adjust.

Keywords: Geoscience education; active learning; video lectures; student assessment

ISBNs: 9781108717847 (PB), 9781108681421 (OC)
ISSNs: 2517-780X (online), 2517–7796 (print)

Contents

1 Introduction

For centuries university students have been listening to lectures from their professors. Clearly that approach can lead to positive outcomes, as students graduate knowing more than when they arrived and they go on to have success in their future careers. But is lecturing the most effective way of helping our students acquire the knowledge and skills they need to be successful scientists? The answer seems to be no (Freeman et al., 2014). Cognitive science research and classroom studies show how learners must actively engage with the material to gain a deep understanding and acquire the skills to successfully apply their knowledge in varied situations (Brown et al., 2014).

Active learning is not a new idea. We have all used it, for example to learn how to play a musical instrument, shoot a basketball, knit a sweater, or to ride a bike or drive a car. We also employ active learning in field trip and lab activities in our courses, where students typically learn to identify fossils by actually looking at fossils. However, active learning exercises still require that the students have some foundational knowledge, including a common vocabulary and familiarity with underlying concepts.

One approach to increase the amount of active learning in the classroom, while recognizing the continuing need to deliver content, is the "flipped classroom" model (for a summary, see Milman, 2012). The flipped classroom gets its name because it inverts the traditional lecture approach (it's also sometimes called an "inverted classroom"), delivering content outside of the classroom and allocating in-class time to active learning exercises. The out-of-class content delivery phase of the learning cycle can take several different forms, and there are similarly many choices for how to structure the in-class activities.

In this Element, I will discuss my experiences at the University of California Santa Cruz (UCSC) in converting three lecture-style courses to flipped classes. I will focus both on benefits and also potential concerns of the flipped class format, and I aim to provide some practical suggestions for overcoming those challenges based on my personal experiences. These courses (Invertebrate Paleobiology, Sedimentology and Stratigraphy, and Statistics and Data Analysis in the Geosciences) are all upper-division electives taken primarily by juniors and seniors, with enrollments in the range of 15–30 students. As a result, my experiences may be different than if the courses were required, had larger enrollment, or were taught at a different academic level. UCSC is a medium-sized and moderately selective public R1 research university. In fall 2016, 42 percent of undergraduates were first-generation college students and

39 percent were Pell Grant recipients (www.universityofcalifornia.edu/infocen ter/fall-enrollment-glance). UCSC is also a Hispanic-Serving Institution and, in fall 2017, 31 percent of Earth and Planetary Science majors were Hispanic or Latino. Our program is large (for an Earth Science department): 198 majors in fall 2017 and on average 60 graduating students per year.

2 Implementing the Flipped Classroom

First, I'll provide some background to describe how I have implemented active learning and converted my courses to a flipped class structure. There is no single "correct" way to flip the classroom so, although I will discuss my choices, it is important to craft policies and procedures that align with your personal teaching style and with the resources and time available to you. Flipping the classroom can go as far as shifting all content delivery to outside of the classroom and spending all in-class time on activities, but that is simply one end of a continuum. You don't have to go from 0 to 100 percent at once. I transitioned the invertebrate paleobiology course to a flipped class with an incremental approach, by initially adding in-class activities to some class meetings, so in those select class meetings we would spend 75 percent of time on lecturing and 25 percent of time on hands-on work. Over time I started recording videos to be able to expand the short activities into hour-long exercises that occupied the entire class time, and started designing more activities for other topics. In contrast, I shifted my sedimentology/stratigraphy course from pure lecture to 100 percent flipped class all at once. For reasons of instructor workload (and sanity) I would not recommend that approach!

To introduce the content, I produced short videos by recording my voice over PowerPoint slides (following a pre-prepared script), using PowerPoint's built-in slideshow recording option. The videos are typically around 10 minutes long, and no more than 15 minutes. I saved the videos as MPEG movie files and then uploaded them to a YouTube channel for the class. These YouTube videos are available for anyone to watch, but it's possible to make the videos unlisted and only available with a link you provide to your students. My approach was fairly low-tech and didn't require video-editing software or any special skills on my part, so it should be a process that anyone can adopt. Having a pre-written script helps the video narration flow more smoothly, but the script could also be posted for students to view. I have not tried this, but there are potential benefits, such as providing a framework for student note-taking, aiding accessibility for international students or English-language learners, and demonstrating the correct spelling of technical terms (where YouTube's automated captioning often fails spectacularly). If you have the desire, there are more sophisticated possibilities for interactive online instruction, but they may require special software or technical support from trained personnel. Although the videos have been surprisingly popular among students, videos or online content are not the only choices for pre-class assignments. To give some examples, you could assign readings from a textbook or the internet, have students read an article from the scientific literature, or ask them to participate in online discussions.

My in-class activities are primarily pen-and-paper worksheets with conceptual exercises, the kind of exercises that many of us assign as homework in a lecture-style course. Some involve hands-on work: For example, students flip coins to generate a bounded random walk, or explore adaptations to soft substrates using metal and clay shapes of different sizes in containers of water-saturated sand (this activity can get a bit messy . . .). Others have students analyze real or simulated data and models on computers during class. The full list of exercises can be found at my website: https://people.ucsc.edu/~mclap ham/eart101.html. I have also found inspiration for activities at the SERC website (Science Education Resource Center; https://serc.carleton.edu), which is a great resource for lessons that can be used or adapted for in-class work. My primary suggestion when designing in-class exercises, such as the design of pre-class videos or readings, is to work within your comfort zone and choose the types of activities that match your teaching style.

The transition to a flipped classroom can be a stepwise process, so it is beneficial to experiment, reflect on what worked and did not work, and then adjust. This advice applies not only to the in-class activities but also to pre-class content delivery. That is not to say the shift is without challenges, but adopting a flipped classroom doesn't require any special training beyond the motivation to try. The challenges associated with the flipped class format can be overcome, and the benefits are well worth the effort.

3 Benefit: Students Learn More

The most fundamental benefit from switching to a flipped class structure that incorporates in-class active learning is that students learn more than they would in a traditional lecture (e.g., Moravec et al., 2010; Deslauriers et al., 2011; Mason et al., 2013; Freeman et al., 2014), although the improvements may derive largely from active learning rather than the flipped structure itself (Jensen et al., 2015). The improved performance from active learning may be especially pronounced for underprepared students (Ryan and Reid, 2016) and those who come from more disadvantaged backgrounds (Haak et al., 2011; Freeman et al., 2014), although not all studies support this (He et al., 2016).

My anecdotal experiences have been consistent with these research findings. Although the flipped classroom isn't a cure-all, it has been noticeable how students do in fact gain a deeper understanding of the material with active learning. I made simultaneous changes to exam rules and structure that prevent a more quantitative assessment, but my qualitative impression is that there has been particular improvement in students' ability to apply their knowledge to more authentic problem-solving situations. The flipped class format allows (more accurately, forces) students to grapple with actual problems, which in many cases can be open-ended and partially ambiguous, providing them with a more realistic experience. Rather than focusing on superficial memorization, students actually gain skills that scientists use. I've been pleasantly surprised by the difference that shifting to a flipped class has made to student confidence and expertise.

3.1 Associated Challenge: Students May Not Recognize That They Are Learning

Because the in-class activities often feature authentic problem-solving tasks, which forces students to confront challenging and unknown questions, students can feel like they are struggling and therefore that they aren't learning. Of course, this is the opposite of the truth! This misperception likely arises because many students don't realize that effortful practice on challenging material and retrieval of information by self-testing actually promote learning (Brown et al., 2014). Instead, people often fall into the illusion that learning occurs when something seems to come easily. A well-presented lecture can make the information seem straightforward, which seduces listeners into believing they have mastered the material (the fluency illusion). There is also sometimes concern among instructors that students' perceptions that they aren't learning as much may lead to more negative student evaluations of teaching. What strategies are there for conveying the benefits of active learning to students?

I'll start by noting that I haven't seen any drop in numerical scores on student evaluations of teaching aside from normal year-to-year fluctuations, and positive written comments about the flipped class format far outweigh negative ones. That said, I have received two instances (over seven flipped class offerings) of strongly negative written comments from students who vehemently disliked the format! I suspect there are also students who dislike the format but who do not complete the student evaluation. I think the concerns about significant negative consequences for instructor promotion/tenure cases are generally overestimated, but I recognize that student reactions may be different depending upon the circumstances and that the criteria by which faculty teaching is reviewed will also differ among institutions. Although my experience has been positive on balance, I have attempted several strategies to further convince students of the benefits.

I began to include a paragraph in the course syllabus that briefly outlines the cognitive science justification for active learning, and I also try to relate instructional design choices back to those principles. If students mention feeling confused or remark that the exercises are difficult, I may emphasize how deliberate and deep thinking is actually a good sign of learning (of course, I also help them clarify their reasoning to move beyond the specific point of confusion). I also try to insert discussions about topics such as the importance of spaced practice, how repeated retrieval of information promotes long-term retention, and how knowledge is more easily acquired when building upon a foundation of prior information. I also emphasize how science, and education more generally, is primarily not about learning a set body of facts, but instead about acquiring the skills to interpret data, solve problems, and apply knowledge in new situations. My impression is that students are generally interested in learning about cognitive and metacognitive strategies, and that these informal discussions do help limit student resistance to active learning.

More recently, I've added questions about student goals and expectations to the questionnaire I distribute on the first day of class. The questionnaire was initially intended to learn about students' preferred name and pronouns and to gauge their relevant prior coursework, but I now also ask two additional questions. The first has the students think about a hobby, sport, or other skill and describe what they did to become proficient at that activity, with the aim of raising the topic of deliberate practice (and sometimes struggle) as an important way of learning. The second asks the students to rank three learning goals in order of personal importance: acquiring information, learning how to use information and knowledge in new situations, and developing lifelong learning skills. Most students rank the latter two goals more highly, which allows me to raise the benefits of active learning in developing skills to apply science as a process rather than as a body of facts. The goal of these prompts is to help prime students to view rigorous effort as a positive trait, so it's possible they lead to some improvement in student attitudes.

4 Benefit: In-Class Work Can Be Made Challenging

The flipped classroom isn't the only way for students to work on exercises; it's also possible to design high-quality homework assignments in a lecture-based course. So why go to the trouble of flipping a class? Although homework assignments also give students opportunities to practice interpreting data or applying the concepts from lecture, I think it's more feasible to implement more open-ended and possibly ambiguous exercises when instructors are on hand to provide guidance. Many of the activities I use in class, especially in the sed/ strat course, are more challenging than I would assign as homework. Working in class with instructors present gives students the confidence to extend themselves and tackle these challenging problems.

4.1 Associated Challenge: Maintaining Student Engagement and Progress

Students only gain the full benefits of active learning and distributed practice if they complete the activities in the sequence and at the time at which they are assigned. However, students have many demands on their time and may not always be able to prioritize your class. How then to best ensure students participate in and are engaged during the in-class work?

I have experimented with several methods of grading to incentivize students to complete the in-class work promptly (Table 1). I initially thought of the classroom activities as in-class equivalents of traditional homework, so I

Table 1 Grading strategies for in-class work

Strategy	Benefits	Drawbacks
Grade based on correctness of answers, exercises turned in at end of class meeting	Incentivizes student effort and promotes timely completion	Highly stressful for students learning material for first time, distracting from learning
Grade based on completion, exercises turned in at any time	Allows students to concentrate on learning and provides flexibility with other demands	Some students let work pile up, negating benefit of spaced practice
Grade based on completion, full credit for exercises turned in by next class, partial credit for at any time later	Provides flexibility when students face other demands, but signals importance of spaced practice	My preferred option, although the incentive for timely completion may not be that strong

graded student work based on performance. Students ideally turned in the exercise at the end of the class meeting, although I sometimes allowed assignments to be turned in at the next class when I misjudged the time that would be required. This approach, in theory, should be very strong motivation for students to attend class and complete the exercises in a timely fashion. However, even though I was deliberately very lenient, assessing performance was still stressful for the students, so it ended up distracting from learning. In hindsight, mixing "teaching" and "testing" was a mistake.

I quickly shifted to grading in-class work for completion only. Assigning points only for completion of the exercise reduced anxiety and allowed students to concentrate on learning, but policies governing the completion of those exercises are still important. Until recently I encouraged students to turn in completed assignments at the end of class or the beginning of the next meeting, but I would give full credit for assignments turned in at any point. The flexibility of this policy was appealing, especially when students had circumstances preventing them from focusing on the class, and I also reasoned that it was better for a student to complete the exercise much later than never at all. However, in rare instances students would fall far behind and then complete a large number of exercises all at once, defeating the purpose of the distributed practice inherent in the course design. After this experimentation I have settled on a compromise approach that assigns full credit for completion of the exercise by the subsequent class and partial credit for completion of the exercise at any time. The difference in points between on-time completion and completion at any time is small enough that the effect on a student's final grade will be minimal, but this system signals the importance of staying on top of the work.

Accommodating the pace at which students work through the exercises is an additional challenge, as some students will pick up the concepts more quickly while others may require more time or will be more thorough in their work. I have found this one of the most difficult logistical issues to address. One possible mitigation is to include additional questions that are not part of the assignment but that students can consider if they have extra time, as a simple form of differentiated instruction. I have built these additional questions into several in-class exercises (although not all of them), and students generally will take the time to work on the extra material. The interactive format also allows me to provide more guidance to students who need support and to let stronger students struggle a bit with the material before I help. I also partly avoid the problem altogether because the in-class exercises tend to run a little long, although that's not ideal. Despite those mitigation strategies, at times some students finish a few minutes early – typically not more than 5 or 10 minutes before the end of the 65-minute class period. If students finish the material early, I don't force them to stay and they're able to leave class. Although this may seem like anathema to professors, and may be problematic if student contact time is monitored by administrators, in principle teaching should focus on student learning rather than time spent.

4.2 Associated Challenge: Accommodating Greater Instructor Workload

Frequent in-class work is valuable for student learning, but the process of giving feedback can impose a significant burden on the instructor's time. I will begin by noting that my teaching load involves 3.25 hours of contact time per week over a 10-week quarter, so strategies that worked for me may not be feasible for someone who teaches more. What strategies are there for maintaining the quality of feedback to students but also reducing the burden on the instructor?

For the first few years of teaching flipped classes, I provided detailed and individualized written feedback on each student's assignments, which hopefully was valuable to them but was extremely time-consuming (1–3 hours of work per assignment) for me! Although feedback is useful in theory, it wasn't clear that students read the comments in any detail, so I have more recently begun to explore other options (Table 2). Delegating the assessment to a teaching assistant or grader could be feasible at some institutions, but I didn't want to add to the workload of our TAs, who already help out during class, teach the lab sections, and come on field trips. It may also be difficult to expect TAs or especially undergraduate graders to provide high-quality feedback, given the range of students' answers to open-ended questions. My latest strategy eliminates individualized feedback and substitutes a detailed key that I make available to the students after the exercises have been submitted. I collect student assignments to scan for widespread misconceptions that we can discuss as a group, but otherwise I don't make any annotations to the students' work. Although this has reduced the assessment workload dramatically, my initial feeling is that this approach is a little too impersonal and that it is also possibly not clear to students whether their answers meet or fall short of expectations. In future class offerings, I plan to adopt a hybrid approach, providing the answer key but making notations (not written comments) on each student's assignment to mark the answers that fail to meet expectations.

There is some initial time commitment to set up the keys, which require more thorough written explanations than would be necessary for my own grading

Table 2 Strategies for feedback on in-class work

Strategy	Benefits	Drawbacks
Instructor provides individualized, written feedback	Granular and detailed information for students	Extremely time-consuming
Teaching assistant provides individualized, written feedback	*Would reduce burden to instructor*	*Teaching assistants likely have high workload already*
Instructor makes answer key available	Detailed information provided to students with little burden on instructor	Impersonal, perhaps not clear to students if their answers meet expectations
Instructor makes answer key available and indicates when student answers fail to meet expectations	Detailed information provided and semi-personalized feedback	None? I think this is the best compromise and it's the approach I'll adopt.

Italics indicate that I have not tried this approach.

purposes, but there will be time-savings over the longer term. This would be especially true for larger classes. However, keys have one potential risk in that they can be passed on to students who take the course in a future offering. If work is entirely in class it is difficult for students to refer to a previous key, but if students are able to complete work outside of class there is a risk that they may instead copy the answers. There is no foolproof way of identifying this sort of academic dishonesty, but answers copied from a key are likely to be unusually similar to the key answers themselves, a good clue especially if the activities are more open-ended or realistic. Providing individualized and detailed written feedback is an aspirational standard, and may be achievable depending on instructor workload and/or availability of teaching assistant resources, but my experience suggests that the use of answer keys is a compromise that maintains most of the educational benefit.

5 Benefit: Students Work More

One lesser-publicized benefit of a flipped classroom, beyond the learning gains from active learning strategies, is that students often must spend more time working on class-related activities. This will not always be the case (e.g., He et al., 2016), depending on the design of any particular class, but in my experience it's rare for a traditional class to assign homework assignments after each meeting. In contrast, the flipped classroom typically requires students to complete a reading or watch a video before every class. For example, I tell students that they should plan to spend 30–45 minutes watching the pre-class video (the video itself is 10–15 minutes), taking good notes, and ensuring that they understand the concepts. Although I don't believe many students actually spend 30–45 minutes, any amount of out-of-class effort is on top of the in-class contact time.

5.1 Associated Challenge: Ensuring Student Compliance with Pre-Class Work

Pre-class content acquisition is an important foundation for the in-class activities, but do students actually complete the assigned pre-class work? The potential that students may not be adequately prepared is a concern for instructors, so what strategies are best for ensuring or incentivizing compliance? Do these strategies motivate students to engage meaningfully with the pre-class work?

First, I'll note that my experience in upper-division elective courses is that students are generally conscientious, although there certainly are times when some students watch the video just as class is beginning or fail to watch the video altogether. The experience may be different for a class that is required or in a general education class designed for non-majors, as some students in those courses may have less motivation for the subject material. That said, I have experimented with different strategies for pre-class compliance (Table 3).

Initially I tried low-stakes online quizzes administered through and automatically graded by our campus learning management system. Most students completed the quizzes, but it's not clear how deeply they engaged with the material because the quizzes were designed to be quite easy. The quizzes also comprised only 10 percent of the overall grade, divided among 20 or more quizzes, so the incentive to complete any single quiz was not that strong. One additional downside was that students would occasionally claim technical difficulties – for example, that they submitted the quiz but the scores didn't appear in their record – and those claims were difficult to adjudicate. It may be less intrusive to integrate assessment questions with the content delivery in a more sophisticated online design, but my use of voice-over PowerPoint videos didn't permit that approach.

I wasn't particularly satisfied with the use of quizzes as a penalty for non-compliance, so I have also experimented with an honor-system approach

Table 3 Strategies for incentivizing compliance on pre-class work

Strategy	Benefits	Drawbacks
Pre-lecture quizzes, but allowing class participation without completion	Signals importance of preparation, so students have foundation for in-class work	Small point value may be weak incentive; may have to adjudicate claims of technical problems
Pre-lecture quizzes, but barring class participation without completion	*Strongest enforcement to ensure students are prepared for in-class work*	*Highly inflexible so doesn't respect that students must balance commitments*
Checking notes in person before/during class	Signals importance of preparation; notes are less superficial than quiz	Logistically impractical; small point value may be weak incentive
Nothing – trust students to make their own decisions	Maximum flexibility for students to balance demands	Some students will not complete pre-class video/reading

Italics indicate that I have not tried this approach.

(although there was still an incentive because students are able to use their notes during the final exam). I also briefly considered but abandoned the idea of assigning a small point value and checking students' notes at the beginning of class, which proved to be a logistical hassle, so I have settled on trusting that students will take charge of their out-of-class learning. There wasn't any noticeable change in student preparation (based on qualitative observation only) after I eliminated the quiz requirement, suggesting that while quizzes likely prompted more students to look at the out-of-class work, those students probably only engaged superficially with the material.

The use of pre-lecture quizzes (or checking of notes) to enforce engagement with out-of-class work raises the issue of what to do with students who still do not complete the tasks. Should those students be barred from participating in the in-class activities because they did not complete the preparatory work? There is a trade-off between allowing students to take charge of their own education, even at the risk of having some students ignore beneficial activities, versus creating requirements to encourage or force students to complete activities that are beneficial to their learning, even if those activities are not the students' highest priority at a particular time. Barring attendance unless students complete the preparatory work should be a strong incentive to students, but is also an inflexible policy that is not accommodating of other demands (school, work, family, health, etc.) on students' time. I have not tried this strategy; in my classes students lost a small number of points for failing to submit a pre-class quiz, but could still participate in class activities. A stricter approach may work for others, but the key is choosing policies that you are comfortable enforcing and that align with your teaching style.

6 Benefit: Greater Interaction with and Among Students

Because students actively work on exercises during the class meeting rather than sitting quietly during a lecture, the flipped class approach provides considerably more opportunities for teacher–student interaction and for group interactions among the students themselves. I find the ability to interact with the students, whether one-on-one or in small groups, to be tremendously rewarding and far more enjoyable than lecturing. Likewise, student surveys consistently rank personal interactions with faculty as an important predictor of satisfaction (Chambliss and Takacs, 2014; Gallup-Purdue Index Report, 2015). Working with students on course-related activities doesn't necessarily foster a close relationship, but I'm still able to get to know students on a more personal level and to talk (briefly, at least) about academic planning, career goals, or issues navigating the university bureaucracy. Even though instructors are committed to student success regardless of the class format, providing individualized guidance as students work through problems also enhances the perception that the classroom is a supportive environment.

The flipped classroom also provides more frequent and more direct opportunities to monitor student learning. Of course, most of us ask questions when lecturing in an attempt to engage students and assess their understanding, but this provides more limited feedback than a conversation where students explain and justify their thought process. The course structure that asks students to turn in written answers for each exercise also allows me to identify and briefly address misconceptions that may be shared among a wider group of students, rather than waiting for the results of infrequent exams. I have found this early assessment to be extremely useful in identifying students who may need more support to succeed in the course, and frequent feedback to students can also help them monitor their own progress. The flipped class also permits differentiation where some students can receive more support and other students can work more independently and even extend themselves to consider more nuanced topics beyond the basics of the lesson.

The interactive format of in-class active learning, often involving group work, has additional benefits for student learning and for career preparation. The process of elaboration, where students explain a concept or their thought process to other students or to the instructor, helps them gain a deeper understanding of the material. My experience has been mixed; in some class offerings the students rarely discuss the questions amongst themselves, despite my encouragement, but in others they are highly active in small groups. This partly seems to depend on the difficulty of the material, with more challenging topics prompting more group work. Group work exposes students to the process of scientific debate, respectfully presenting their interpretations and listening to the opinions of others to reach better conclusions. Even if the exercises are not group projects in the strict sense, students become more skilled at group interactions and learn how to be effective team players, assets that are highly valued by employers and in academia. Finally, group work helps the students to get to know each other, which helps to build a stronger and more supportive community within the department itself.

6.1 Associated Challenge: Personnel Requirements and Logistics

Assigning in-class activities and providing sufficient guidance and support to students during that in-class work requires a small student-to-instructor ratio. There are other types of active learning that can be accomplished in larger classes (think-pair-share, one-minute papers, etc.), but the activities that I have implemented seem like they would be difficult to achieve with a student-to-instructor ratio greater than 15:1 or 20:1. For challenging topics, eliciting more frequent student questions and requiring closer guidance, a ratio closer to 10:1 would be ideal. I have been fortunate that our department long ago made the decision to prioritize graduate teaching assistant resources in a way that supports this kind of intensive teaching (at the expense of TA resources in our large general education courses). For various reasons, you may not have access to as much instructional support, so what can be done in those situations? Aside from longer-term options like lobbying your department to shift how teaching assistants are allocated, or unlikely possibilities such as receiving more TA resources from your administration, recruiting past undergraduates to help out in exchange for course credit is a strategy that I have considered. Many schools, in particular undergraduate-only institutions, already deploy undergraduates as course assistants. If it is not feasible to increase the number of instructors (including graduate and undergraduate student assistants), reducing the complexity or ambiguity of the exercises can mitigate the need for frequent guidance. One of the benefits of guided in-class active learning is the ability to incorporate more realistic and open-ended problems, but those problems also tend to require more hand-holding. Students can still receive much of the benefit of active learning with more concrete and prescribed activities. Finally, you can always just implement activities with the knowledge that you won't be able to interact with each student as often or in as much detail as you ideally would like. In many cases we could provide richer experiences if additional resources were available, so don't let perfect be the enemy of good!

The physical layout of the classroom is another potential logistical challenge, at least for larger classes. Active learning exercises in which the students work in small groups, and the instructors circulate to discuss topics with the students, need rooms where instructors and students can easily mingle. This is not an easy task in the typical lecture hall, with its rows of seats! Fortunately, many of our departments have teaching labs with tables or benches that can accommodate 20–40 students. For most paleontology courses, and Earth science courses in general, that capacity is adequate to run any kind of active-learning exercise. If you teach a larger course, however, are there strategies to overcome the challenge of room layouts? Perhaps not much can be done quickly, or on a personal or department level. Active learning is becoming more widespread, so campuses are beginning to consider the requirements of in-class work when renovating classrooms, so you could lobby your campus administration. The more immediate solution, if all available rooms have a lecture-theater layout, is to implement different kinds of active learning more feasible for larger lecture classrooms.

7 Conclusions

The flipped classroom has numerous benefits, most notably improved student learning but also more rewarding interactions between instructors and students. In my experience teaching flipped classes, students become more confident scientists and are better able to apply their knowledge to realistic problems, important skills as they move to the next phase of their careers. There are logistical challenges associated with converting a course to a flipped class, but those challenges are less severe than one might expect and can easily be overcome. Student resistance to active learning is rare, although not absent, but can be mitigated by introducing students to the cognitive science basis of learning. In-class activities are the critical component, so it is also crucial to ensure student engagement. Grading assignments for completion, with full credit for timely completion and partial credit for completion at any time, creates a balance between emphasizing the importance of the in-class work and respecting the many demands on student lives. Giving individualized feedback can be extremely time-consuming for instructors, but that commitment can be minimized while maintaining the quality of feedback through the sharing of answer keys. Ensuring student compliance with pre-class assignments is the most difficult logistical challenge. Quizzes to assess completion are quite feasible to implement. However, I have been satisfied with trusting students to make their own choices, even though not all students complete the assignments. The personal interaction with students is the most extremely rewarding aspect of the flipped class, but is only feasible with small student-to-instructor ratios and classrooms physically designed to permit interaction. These resource challenges can be the hardest to overcome because they are largely beyond the instructor's control, but the activities can be tailored to fit within the resource constraints. Converting a course to use a flipped class structure doesn't require any special skills or training, so it can be successfully implemented by anyone with time and interest. Most importantly, try out different approaches, reflect on their success, and continually make adjustments. The effort to design a flipped course will pay dividends through enhanced learning and improved student success.

References

Brown, P. C., Roediger III, H. L., & McDaniel, M. A. (2014). Make It Stick: The Science of Successful Learning. Cambridge, MA: Harvard University Press.

Chambliss, D. F. & Takacs, C. G. (2014). *How College Works*. Cambridge, MA: Harvard University Press.

Deslauriers, L., Schelew, E. & Wieman, C. (2011). Improved learning in a large-enrollment physics class. *Science*, **332**, 862–4.

Freeman, S., Eddy, S. L., McDonough, M., Smith, M. K., Okoroafor, N., Jordt, H. & Wenderoth, M. P. (2014). Active learning increases student performance in science, engineering, and mathematics. *Proceedings of the National Academy of Sciences, USA*, **111**, 8410–5.

Gallup-Purdue Index Report. (2015). Retrieved November 26, 2017, www .gallup.com/services/185924/gallup-purdue-index-2015-report.aspx.

Haak, D. C., HilleRisLambers, J., Pitre, E. & Freeman, S. (2011). Increased structure and active learning reduce the achievement gap in introductory biology. *Science*, **332**, 1213–6.

He, W. L., Holton, A., Farkas, G. & Warschauer, M. (2016). The effects of flipped instruction on out-of-class study time, exam performance, and student perceptions. *Learning and Instruction*, **45**, 61–71.

Jensen, J. L., Kummer, T. A. & Godoy, P. D. d. M. (2015). Improvements from a flipped classroom may simply be the fruits of active learning. *CBE—Life Sciences Education*, **14**, 1–12.

Mason, G. S., Shuman, T. R. & Cook, K. E. (2013). Comparing the effectiveness of an inverted classroom to a traditional classroom in an upper-division engineering course. *IEEE Transactions on Education*, **56**, 430–5.

Milman, N. (2012). The flipped classroom strategy: What is it and how can it best be used? *Distance Learning*, **9**(3), 85–7.

Moravec, M., Williams, A., Aguilar-Roca, N. & O'Dowd, D. K. (2010). Learn before lecture: a strategy that improves learning outcomes in a large introductory biology class. *CBE—Life Sciences Education*, **9**, 473–81.

Ryan, M. D. & Reid, S. A. (2016). Impact of the flipped classroom on student performance and retention: a parallel controlled study in general chemistry. *Journal of Chemical Education*, **93**, 13–23.

Acknowledgments

Thanks to Susan Schwartz for discussions and ideas that arose from our department newsletter article on this topic. Alycia Stigall and an anonymous reviewer provided thoughtful feedback.

Cambridge Elements ≡

Elements of Paleontology

Editor-in-Chief

Colin D. Sumrall
University of Tennessee

About the Series

The Elements of Paleontology series is a publishing collaboration between the Paleontological Society and Cambridge University Press. The series covers the full spectrum of topics in paleontology and paleobiology, and related topics in the Earth and life sciences of interest to students and researchers of paleontology.

The Paleontological Society is an international nonprofit organization devoted exclusively to the science of paleontology: invertebrate and vertebrate paleontology, micropaleontology, and paleobotany. The Society's mission is to advance the study of the fossil record through scientific research, education, and advocacy. Its vision is to be a leading global advocate for understanding life's history and evolution. The Society has several membership categories, including regular, amateur/avocational, student, and retired. Members, representing some 40 countries, include professional paleontologists, academicians, science editors, Earth science teachers, museum specialists, undergraduate and graduate students, postdoctoral scholars, and amateur/avocational paleontologists.

Paleontological
S O C I E T Y

Cambridge Elements ≡

Elements of Paleontology

Elements in the Series

These Elements are contributions to the Paleontological Short Course on *Pedagogy and Technology in the Modern Paleontology Classroom* (organized by Phoebe A. Cohen, Rowan Lockwood, and Lisa Boush), convened at the Geological Society of America Annual Meeting in November 2018 (Indianapolis, Indiana USA).

Flipping the Paleontology Classroom: Benefits, Challenges, and Strategies
Matthew E. Clapham

Integrating Macrostrat and Rockd into Undergraduate Earth Science Teaching
Phoebe A. Cohen, Rowan Lockwood, and Shanan Peters

Student-Centered Teaching in Paleontology and Geoscience Classrooms
Robyn Mieko Dahl

Beyond Hands On: Incorporating Kinesthetic Learning in an Undergraduate Paleontology Class
David W. Goldsmith

Incorporating Research into Undergraduate Paleontology Courses: Or a Tale of 23,276 Mulinia
Patricia H. Kelley

Utilizing the Paleobiology Database to Provide Educational Opportunities for Undergraduates
Rowan Lockwood, Phoebe A. Cohen, Mark D. Uhen, and Katherine Ryker

Integrating Active Learning into Paleontology Classes
Alison N. Olcott

Dinosaurs: A Catalyst for Critical Thought
Darrin Pagnac

Confronting Prior Conceptions in Paleontology Courses
Margaret M. Yacobucci

The Neotoma Paleoecology Database: A Research Outreach Nexus
Simon J. Goring, Russell Graham, Shane Oeffler, Amy Myrbo, James S. Oliver, Carol Ormond, and John W. Williams

Equity, Culture, and Place in Teaching Paleontology: Student-Centered Pedagogy for Broadening Participation
Christy C. Visaggi

A full series listing is available at: www.cambridge.org/EPLY

For EU product safety concerns, contact us at Calle de José Abascal, 56–1°,
28003 Madrid, Spain or eugpsr@cambridge.org.

www.ingramcontent.com/pod-product-compliance
Ingram Content Group UK Ltd.
Pitfield, Milton Keynes, MK11 3LW, UK
UKHW040945090126
466816UK00019B/295